Fun with Materials

written by Maria Gordon
and
illustrated by Mike Gordon

RSVP

RAINTREE
STECK-VAUGHN
PUBLISHERS
The Steck-Vaughn Company

Austin, Texas

Simple Science

Day and Night

Electricity and Magnetism

Float and Sink

Fun with Color

Fun with Heat

Fun with Light

Fun with Materials

Push and Pull

Rocks and Soil

Skeletons and Movement

Published by Raintree Steck-Vaughn Publishers, an imprint of Steck-Vaughn Company

Library of Congress Cataloging-in-Publication Data
Fun with materials / written by Maria Gordon and illustrated by Mike Gordon.
   p.   cm.—(Simple science)
   Includes index.
   Summary: An introduction to the materials in the world around us, with suggestions for activities for exploring their properties.
   ISBN 0-8172-4505-7
   1. Materials—Juvenile literature.
   2. Materials—Experiments—Juvenile literature.
   [1. Materials—Experiments. 2. Experiments.]
   I. Gordon, Mike, ill. II. Title. III. Series: Gordon, Maria. Simple science.
   TA403.2.G67   1996
   620.1'1—dc20   95-25397

Printed in Italy
1 2 3 4 5 6 7 8 9 0 00 99 98 97 96

Contents

Touch and look at some things around you. They are made of different things. The things they are made of are called materials.

Metal, wood, and cotton are materials.
Can you find...

a saucepan made
of metal...

a table made of wood...

a T-shirt made
of cotton?

A metal saucepan feels hard. It doesn't burn. A wooden table feels hard. It holds things up. A cotton T-shirt is soft. It is comfortable and it keeps you warm.

Different materials are good for different things.

Would you like to wear a metal T-shirt...

or cook with a wooden saucepan?

Cave people made arrowheads and axes from stones. They used clay to make pots and powder from rocks to make paint. Huts were built with mud and parts of plants.

People learned to weave leaves and stems to make baskets, ropes, and nets. More and more materials were used to make different things. There were statues made of stone, pens made of feathers, boats made of wood, and horseshoes made of iron.

Pour some water.

Watch it spread out.

Put your finger in it.

Now pull it out.

It does not leave a hole.
Water is liquid.

If you put a stone into a cup, it does not spread out.

But if something makes a hole in a stone, the hole does not go away. Stones are solid.

Half fill a plastic bag with water.

The water stays in the bottom of the bag.

Take another plastic bag and blow it up. The air does not stay at the bottom. It fills the bag.

Air is a gas. All materials are solids, liquids, or gases.

Gases can be useful materials. If you whip air into cream, the air makes it frothy.

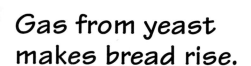

Gas from yeast makes bread rise.

If you switch on a fluorescent light, gas inside the tube glows and helps you to see.

Many solids are useful materials. Solids keep their shape and size.

You can build with solid building blocks. They don't float up in the air! They don't mix with the floor! Tables don't spread out and fill the room, and paper doesn't pour away!

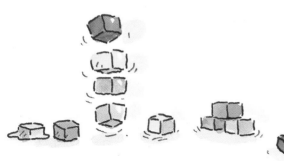

Liquids often mix with other materials. Add water to sand. The water makes the sand stiff. Now you can make sandcastles.

You can make things by mixing solids, liquids, and gases together. Ask an adult to help you make a milkshake. Use…

1 cup of milk
1 cup of vanilla ice cream
2 sliced bananas
1 carton of fresh strawberries

Wash and cut the tops off the strawberries. Put all the solids and the milk into a blender. Put the cover on the blender, switch it on, and watch air mix in and froth up your shake!

Look for solid,
liquid, and gas materials.

What do these materials look like?
What colors are they?
How do they feel?

Touch a new piece
of aluminum foil.
It is hard and
smooth. It looks
shiny.

Look at sand. It
is made of tiny,
hard pieces.
It feels gritty.

Pour some flour.
It is made of even tinier
pieces. It is very soft.

Feel some wool.
It is soft and
light. What
color is it?

Touch some
honey. Is it soft
and runny or
lumpy and stiff?

YUM,
YUM!

Materials can change when you *do* things to them. Crumple some aluminum foil. Now it feels rough.

Pour water from high up. Watch it move fast. Bend and twist some modeling clay. Feel it *go* soft.

Ask an adult to help you with these projects.

Stir a spoonful of sugar into a glass of hot water. Watch the sugar dissolve.

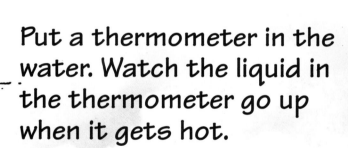

Put a thermometer in the water. Watch the liquid in the thermometer go up when it gets hot.

Heat ice cubes in a saucepan. Watch them change into water. Heat the water until it bubbles. Watch the water change into steam. Changing materials makes them useful for different things.

Materials come from the world around us.
Wood comes from trees.
Sand and oil are found in rocks.

Rubber and cotton come from plants.

Wool and leather come from animals.

Fur coats also come from animals.

These sorts of materials are called natural materials.

Some natural materials can be mixed to make new ones.

Add water to flour. This makes a sticky paste. Soak newspaper in the paste. It becomes stiff and hard as it dries. This is papier mâché.

Scientists have learned how to change natural materials. They make new materials that are not found in the world around us. These are called artificial materials.

Look at a doll with moveable arms and a telephone. Today, these are made of plastic.

Feel a pair of stockings. These are made of nylon. Nylon and plastic are artificial materials.

23

Some things make materials change. Find a rusty nail. The nail is made of iron, but it has mixed with air and water and made rust. Feel an old tree stump. Rain can make wood soft and crumbly. Look at bricks in a wall. Wind and rain can wear away the cement between the bricks.

Some insects and other animals can make materials.

Bees make honey and wax.

Silkworms make silk.

Wasps chew up wood and plants to make papery nests.

What materials were used to build your house? Why were they used? Would you make windows out of plastic wrap or glass? Why? Would you make a roof out of cardboard or tiles? Why? Would you make doors out of stone or wood? Why?

Over the years, people have used different kinds of materials to build houses. Different materials are used in different countries too. In hot countries, windows can be holes. In cold countries, glass makes strong, clear windows. Roofs made of parts of plants can keep out rain, but shingles last longer. Stone doors are very heavy. Wood is easier to cut and use. In places where there is very little rain, houses can be made of dried mud.

The world keeps on making materials.
It uses old materials to make new ones.
This is called recycling. Old trees and plants
crumble into soil. Soil helps grow new
trees and plants.

People must also recycle
because we are using up
materials faster than
the world is making them.

Look at the next
page. What
materials can
you see? Which
ones can you
recycle? The
answers are on
page 31.

Additional projects

Here are a few more projects to test out materials. The projects go with the pages listed next to them. These projects are harder than the ones in the book, so be sure to ask an adult to help you.

4/5	Make a list of the things in your classroom, bedroom, or kitchen. Try to write down the materials that every item is made of.
6/7	Make boats out of paper, lollipop sticks, food containers, bottlecaps, etc. Note which materials you used. Predict which ones will sail the longest, farthest, and fastest and test them on a pond or pool.
8/9	Visit museums. Try weaving. Play with clay.
10/11	Pour, freeze, and boil water. Water is the only substance on Earth naturally found as a gas, a liquid, and a solid. Talk about which things around you are solids, liquids, and gases.
12/13	Make bread that uses yeast.
16/17	List the different qualities of solids, liquids, and gases (rough, sticky, cold, etc.). What qualities do liquids and gases have in common? How are they different? Test materials for strength and insulation.
18/19	Visit a foundry or a glass works. Watch someone make pottery.
20/21	Display raw materials. Research their value. Find out what countries export these materials. Find out how Amazon peoples waterproofed their clothing and bodies with rubber plant sap, how Egyptians made paper from papyrus, etc. See how dandelion sap, a natural latex, dries.
22/23	Make papier mâché masks and other items.
24/25	Spot signs of weathering on buildings and sidewalks. Look at erosion of rocks.

26/27   Read "The Three Little Pigs" and write a new version where the pigs use other materials, including modern building materials like concrete and plastic.

28/29   Join a recycling project in your community. Visit a dump and a recycling plant.

Answers

All these materials can be recycled. Check with your local recycling program.

1. Paper
2. Plastic bottles
3. Cotton and thread
4. Nails and other scrap metal
5. Wood
6. Water
7. Aluminum and steel cans
8. Glass jars and bottles

Other books to read

Amos, Janine. **Waste and Recycling.** First Starts. Milwaukee: Raintree Steck-Vaughn, 1992.

Mitgutsch, Ali. **From Ore to Spoon.** Start to Finish. Minneapolis: Carolrhoda Books, 1981.

Mitgutsch, Ali. **From Tree to Table.** Start to Finish. Minneapolis: Carolrhoda Books, 1981.

Reymond, Jean-Pierre. **Metals: Born of Earth and Fire.** Young Discovery Library. Chicago: Childrens Press, 1988.

Rowe, Julian and Perham, Molly. **Feel and Touch.** First Science. Chicago: Childrens Press, 1993.

Index